How Do You Stay Well?

by Caroline Hutchinson

I0892513

How do you stay well?
There are many things
that you can do.

You can eat many kinds of food to stay well. Do you eat salad? Do you eat fruit? These are some foods that can help you stay well.

You can exercise to stay well, too. Do you like to skate? Do you like to bike? These are some ways to get exercise.

To stay well, you need to keep your body clean.
Do you wash your hands?
Do you take a bath every day?

You need to keep
your teeth clean, too.
Do you brush your teeth
after eating?

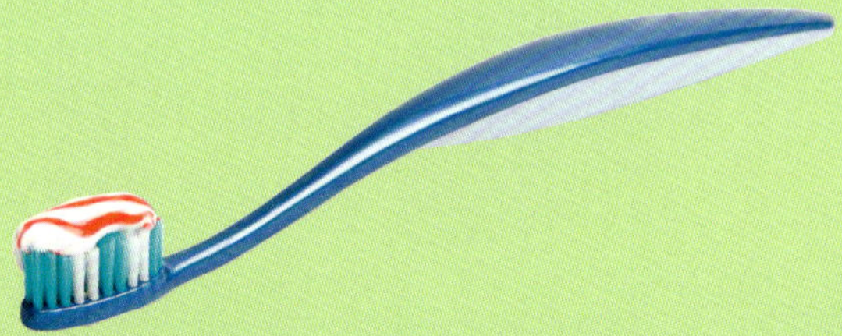

You can stay well by getting the sleep you need. Your body needs about eight hours of sleep every night. How much sleep do you get?

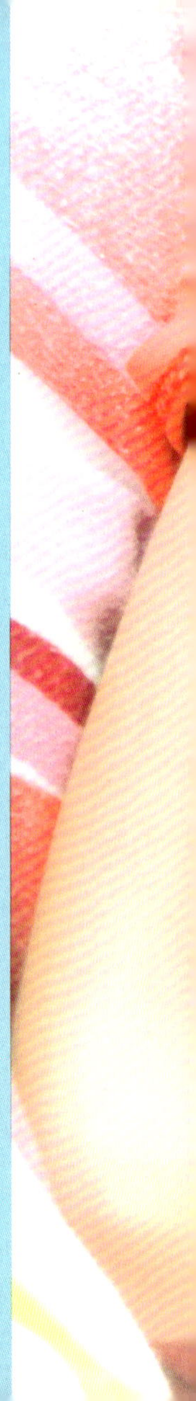

Your doctor helps you stay well. She can help you get well if you are sick.

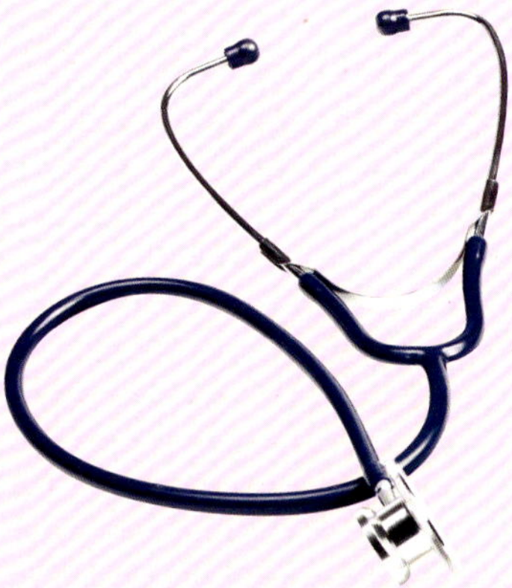

Take good care of your body.
You can't live without it!